P. C. Majumdar

Therapeutics of Cholera Asiatica

P. C. Majumdar

Therapeutics of Cholera Asiatica

ISBN/EAN: 9783337320522

Hergestellt in Europa, USA, Kanada, Australien, Japan

Cover: Foto ©berggeist007 / pixelio.de

Weitere Bücher finden Sie auf **www.hansebooks.com**

THERAPEUTICS

OF

CHOLERA.

(CHOLERA ASIATICA.)

BY

P. C. MAJUMDÁR, M. D.,

Graduate of Medical College, Calcutta, India, Corresponding
Member of the American Institute of Homœopathy, and
Honorary Member of the International Hahne-
mannian Association, etc.

PHILADELPHIA :
BOERICKE & TAFEL.
1893.

PREFACE.

These few pages on the therapeutics of cholera are offered to the profession as the fruit of my continual labors for fourteen years in the so-called " home of cholera." The disputed and doubtful questions of various theories as to the etiology of the disease and its pathological speculations are purposely left out of consideration. The practical points on prevention of cholera, and the means of combatting it when actual invasion takes place, are dealt with in full. The symptomatic indications are given at the end of the book.

The first portion of this book was read to the World's Homœopathic Congress, held at Chicago, in connection with the World's Columbian Exposition, 1893.

Many valuable hints are taken from the works of Hahnemann, Hering, Dunham, Bell, Farrington, H. C. Allen, L. Salzer, G. H. Clark, and others, for which my grateful acknowledgments are due.

P. C. MAJUMDAR.

Philadelphia, Pa., June 12, 1893.

THERAPEUTICS OF CHOLERA.

(CHOLERA ASIATICA.)

Cholera is a most dreadful and fatal disease. It is characterized by purging, vomiting, pains in the abdomen, cramps in the extremities, coldness of body, profuse perspiration, loss of pulse, suppression of urine, great prostration, feeble and hoarse voice, difficulty of breathing, and other symptoms of collapse. Death generally takes place very rapidly, but often it takes on a protracted course. It is a specific epidemic disease, but is endemic in India.

How powerless is all the Allopathic treatment to grapple with the disease! On the other hand, the Homœopathic system is eminently successful in this disease.

PREVENTIVE HYGIENIC MEASURES.

As prevention is better than cure, our sole aim would be to prevent the onset of this dreadful disease, and to minimize its ravages. A great deal has been written on the subject, and numerous methods have been adopted from time to time. In India the great majority of cholera cases take their origin from defective hygienic arrangements. The people of the country take very little care with regard to food, exercise, ventilation, and personal and general cleanliness. Neglect of these measures often is the cause of a cholera outbreak. Many times the houses are so built as to cause a great hindrance to free ventilation of air. These ill built houses are, moreover, overcrowded with a large number of persons. We are all aware how dangerous are the effects of such overcrowding. It is not only the poor people who are obliged to lead such a life, but even rich people also are utterly careless in these important subjects.

Cholera is a preventable disease to a great extent. Hygienic measures are of great importance as preventives of cholera. Calcutta, the metropolis of India, and a big town, was formerly visited by epidemics of cholera, but it has been surprisingly free from it for the last three or four years. It is simply owing to the fact that our municipality is taking particular care about free drainage, ventilation and pure supply of water.

Purity of food and drink, cleanliness of person, free ventilation of houses, daily exercise in open air, constant healthy occupation and avoidance of all dread of the disease will tend to the maintenance of a degree of health, enabling most persons to resist the possible attack of cholera.

"Among the precautionary measures which it behooves each individual to observe, the following are the most important: As much as possible a regular mode of living, the use of simple and substantial nourishment, the slightest possible deviation from one's regular mode of living, avoidance of all debilitating influences, such as excessive mental excitement, depressing emotions, more particularly avoidance of excessive use of spirits." (Baehr.)

Persons must take care not to get chilled or overheated, which are very injurious. I know persons who have an idea that cholera takes place from heat, and so they advise cold bathing in the beginning of an attack. These cases are generally fatal.

All depressing passions and emotions should be carefully avoided. Violent and long-continued exercise is hurtful. Fasting and intemperance in eating and drinking are reckoned as the prolific sources of cholera outbreaks. Use of vegetable food and abstinence from animal food are regarded by competent authorities as the causes of cholera. This is a very doubtful question. Food should be nourishing, and at the same time easily digestible. For this purpose animal food is preferable, but putrid and decomposing fish and meat are very harmful.

Every precaution is necessary to avoid overloading and oppressing the stomach and digestive organs generally. Excessive quantity of anything is bad.

These are sanitary preventive measures; scrupulous attention to these is likely to be rewarded with perfect freedom from an invasion of cholera and other serious diseases of like nature.

It must not be concealed that there are some outbreaks of cholera the cause of which cannot be traced to the unsanitary condition of the place. We have often seen the sudden outbreaks of cholera without any seeming reference to these unsanitary conditions of the place. These are, as it is often said, the miasmatic influences or general atmosphoric changes. We cannot sufficiently be on our guard to prevent this kind of epidemic invasion. But so much can be be assured from personal experience, that even this careful attention to food, drink and cleanliness is sufficient to prevent the spread of the disease.

PROPHYLACTICS.

I may briefly mention here the prophylactic medicinal treatment of cholera. I must confess that my experience is very limited in this direction. *Veratrum*, *Cuprum* and *Camphor* are vouched for as the best preventives of cholera. Hahnemann was the first

physician who discovered the prophylactic virtues of these remedial agents.
In his later writings he wrote about the
preventive power of *Cuprum* in the following words: " The above preparation
of copper, together with good and mod-
erate diet and proper attention to cleanliness is the most certain preventive
and protective remedy." The value of
copper as a preventive is borne out by
some facts in India. Here the lower
class of people use copper coin next to
their skin for this purpose. Our
learned colleague, Dr. Mohendu Lal
Sirca mentioned this fact: "On the
authority of a deputy inspector of
schools, he cited an instance where the
head master of a school in a district
where cholera frequently prevailed, in-
duced a number of people to wear a
pice (a copper coin) through which a
hole was bored and a thread inserted to
fasten it round the body. During five
years, two epidemics occurred, and not

one person wearing the *pice* was affected. That the workers of copper mines are said to be peculiarly free from the ravages of cholera is an almost well established fact. Some urge that a dose of *Veratrum* in the morning and a similar dose of *Cuprum met.* in the evening are sure to protect a person from an attack of cholera. *Camphor*, internally taken or by inhalation is said to be a prophylactic against the disease. The people of India have a very strong faith in *Camphor*, as they say that *Camphor* has the power of destroying the cholera miasm. I have given an extensive trial of this medicine, and with good result. I myself take an inhalation of *Camphor* solution when visiting cholera patients. Dr. Hering's powder of sulphur in the stocking during cholera epidemic is said to be a very valuable preventive. The burning of sulphur and resin has become a household duty in every Indian house. The people have a conviction that these agents have the power of destroying the germs of many zymotic diseases, and thereby purifying the atmosphere.

Whatever may be the intrinsic merit of these remedial measures as a prophylactic of cholera, one fact must always be borne in mind: that doing something in the shape of a preventive during an epidemic of cholera has a great influence on the minds of the people. We are all aware how panic-stricken has become the public mind in the face of a cholera epidemic, and we often see an actual attack from a mere *dread* of the disease. In these cases previous administration of a preventive almost invariably has the desired effect.

CURATIVE TREATMENT.

For convenience in describing the medicines in the treatment of cholera, it is usual with authors to divide the disease into various stages. It is not

exactly that these stages appear one
after the other in regular succession as
described, but on the contrary we often
find one stage merging into the other.
It cannot be expected to see the actual
disease developed as described in the
books. All writers on cholera agree in
recognizing the following stages of the
disease. First, the premonitory stage ;
second, stage of evacuation or full development; third, collapse stage, and,
fourth, the stage of reaction. We shall
describe the treatment in this order, re-
viewing the complications and sequelæ
of the disease at the end.

The number of remedies in true cholera is not very great. Hahnemann first
suggested *Camphor*, *Veratrum alb.* and
Cuprum, and this suggestion is so simply recorded that there is no difficulty
in treating the disease.

In the first stage, it is difficult to
recognize the true nature of the dis-

ease, so people pay very little attention to it; the physician is not generally called at this stage. But if there are malaise, great prostration, pains in various parts of the body, and other uneasy sensations, a few doses of *Aconite* will remove them If there are digestive derangements—loathing of food, no appetite, thin, diarrhœic stools, *Nux vomica*, *Pulsatilla* and *Veratrum* may be useful. In premonitory diarrhœa I generally use either *Camphor* or *Veratrum* after each stool, and that is generally sufficient to prevent the further development of the disease.

In the stage of purging and vomiting, or, in other words, when the evacuations fully set in, the following medicines are to be considered:

Veratrum alb. {
  Camphor.
  Cuprum met. or *acet.*
  Ricinis.
  Jatropha.
  Euphorbia.
  Croton tig.
  Antimonium tart.
  Elaterium.
}

Veratrum may be considered as the type of a class of remedies which are more or less potent in checking an undue evacuation and bringing it into natural color and consistency. In fact, by the administration of one of these remedies according to indications, the further mischief may be averted.

Veratrum album.—As students of the old school, we are very familiar with the fact that *Veratrum* is a drastic purgative, so, according to the Homœopathic law of cure, it must be a medicine *par excellence* for choleraic evacuations — both purging and vomiting. From our repeated experience, we can give *Vera-*

trum the highest place in the developed stage of cholera. The late lamented Professor Farrington says: "*Veratrum* seems to act prominently on the abdominal organs, acting probably through the splanchnic nerves. When these nerves are paralyzed, the blood vessels become overcharged with blood, and pour forth their serum. The prostration, the coldness, the terrible sinking sensation that belong to *Veratrum*, all start from these nerves."

Indications for administering *Veratrum :* Vomiting and purging of a large quantity of serous fluid—"rice water" evacuations as they are called ; colicky pains through the abdomen ; cramps in the extremities, especially in the calves of the legs ; great prostration ; cold sweat, especially on the forehead ; coldness and blueness of the face and hands ; great thirst for large quantities of cold water and acid drinks.

In times of cholera outbreaks it is wise to give *Veratrum* at the first appearance of diarrhœa, so that no further and serious development would take place. In such cases *Veratrum* has marvelous effects. We have many a times saved numbers of cases by the timely administration of this remedy. It is true that in *Veratrum* poisoning, the stools are not always choleric in nature; they are distinctly bilious, greenish, watery, with flakes, and there may not be total suppression of urine, but whatever may be the toxicological effect of *Veratrum* about the evacuations, our clinical experience is very wide as regards the curative results concerned. So we can confidently advise its use in all kinds of stools and vomiting.

In cholera, general depression of strength is very great, and here *Veratrum* is also our sheet anchor. Hahnemann gives the following symptoms of poisoning in his "Lesser Writings:"

"Two children took *White Hellebore* by mistake. A few minutes after taking the drug they became quite cold, fell down, their eyes projecting like those of a person in a state of suffocation; the saliva ran continually from their mouths, and they seemed devoid of consciousness. I saw them half an hour after the accident, and when I arrived both seemed at the point of death; distorted, projecting eyes; disfigured, cold countenance; relaxed muscles; closed jaws; imperceptible respiration."

As regards dose, I generally commence with the 12x and subsequently to 30x. *Veratrum* 3d or 6th centesimal trituration is sometimes useful.

Our next great anti-choleraic remedy is *Camphor*. It is used in the preliminary diarrhœic stage, as well as in the collapse stage. When the system is overwhelmed with the cholera poison *Camphor* should be thought of. The

body is icy cold, voice husky and prostration is intense. As soon as the patient is passing diarrhœic stools no
time should be lost in administering
the remedy. In this stage, if we give
from one to five drops of *Camphor* after
each stool we are almost sure of checking the further progress of the disease.
Hahnemann says: "In the first stage
Camphor gives rapid relief, but the patient's friends must themselves employ
it, as this stage soon ends either in
death or in the second stage which is
more difficult to be cured, and not with
Camphor. In the first stage, accord-
ingly, the patient must get as often as
possible (at least every five minutes)
a drop of *Spirit of Camphor* (made
with one ounce of *Camphor* to twelve
of alcohol) on a lump of sugar or in a
spoonful of water.

"The quicker all this is done at the
first onset of the first stage of the disease, the more rapidly and certainly

will the patient recover, often in a couple of hours; warmth, strength, consciousness, rest and sleep return, and he is saved."

What Hahnemann said above has all been very well proved in our own practice in India. In my younger days, when I was called in the very beginning of a cholera attack, I was almost invariably successful with *Camphor* alone, but later on I have scarcely a case of cholera where I got the opportunity of administering *Camphor*, as the stage is advanced and *Camphor* has no place.

Cuprum.— It is really a very efficacious remedy in the developed stage of cholera. It generally checks purging and vomiting, and is pre-eminently useful in cutting short the distressing and painful cramps in the extremities and other parts of the body. Hahnemann placed great confidence in this remedy.

He sometimes advised us to give it in alternation with *Veratrum*. Our late lamented Dr. Bhaduri, who had treated more cases of cholera than anybody in Calcutta, used to say that he could treat almost all his cases with *Cuprum* alone. He was very fond of *Cuprum arsen.* in the stage of collapse with purging, vomiting and cramps. Drs. Drysdale and Russel, of England, speak very highly of *Cuprum*, as does Mr. Proctor also. This latter gentleman treated ninety-eight fully developed cases of cholera with this drug, and was satisfied with it. He writes: "For the cramps it was unquestionably the best remedy, and I may say for vomiting also. In the stage of collapse I gradually found myself trusting to *Cuprum*, and the impression is very strong in my mind that in collapse it is the most reliable of our remedies."

Indications for *Cuprum:* Purging and vomiting of rice-water fluid; colic of a paroxysmal nature; constant restlessness; cramps in the extremities, beginning in fingers and toes; great exhaustion; spasms in calves and abdomen; icy coldness of the hands and feet; quick, rattling and short breathing; almost imperceptible, weak and thready pulse; pale and sunken features; great thirst—the water runs down with a gurgling noise; relief of vomiting, after drinking; urine scanty or entirely suppressed.

I prefer the higher dilutions, commencing with the 12x and going higher. I have seen a distinct aggravation from the 3d and 6th. *Cuprum ars.* is very useful in cholera. Dr. E. M. Hale first drew our attention to its use in cases of cholera. He says: "I first used it in some severe cases of cholera in the year 1867 and 1876. These cases were marked by the usual intestinal disorders, to which were added severe and painful cramps in the abdomen and ex-

tremities. The alternation of *Arsenic*
and *Cuprum* did not prove as satisfactory as I expected, but the use of *Cuprum ars.* in the 6th trituration, in water for children and dry on the tongue
in adults, generally acted promptly. I
can recommend it in cholera infantum,
spasmodic and neuralgic pains in the
bowels, accompanied by screams and
cramps in the fingers and toes, attended
with great debility and threatened collapse. I have used it in several cases,
and can bear testimony of its value in
cases indicated. Allied to *Veratrum*
and *Camphor* are a number of medi-
icines more or less applicable to check
choleraic evacuations, and these are
*Ricinus, Jatropha, Euphorbium, Croton
tig., Elaterium* and *Antimon. tart.*
Among these we had very satisfactory
‚results from *Ricinus* in the epidemic of
1883.

Ricinus is used in cases of diarrhœic
cholera. I mean cases which assume

the nature of true cholera from indigestion or simple diarrhœa. We had numbers of cases reported in the *Indian Homœopathic Review* of that year. In a large family in Calcutta, there was an outbreak of cholera, and three persons died of it, notwithstanding Homœopathic treatment was resorted to from the beginning. I was called when a fourth case appeared, and at once gave *Ricinus*, which had marvelous effect in restoring the patient to health. There were four more cases in this family, and they were all saved by the timely administration of this medicine. In this house I met a medical student who watched all the cases and was struck by the prompt action of my medicine. He was curious to know the name of the medicine. I told him it was *Ricinus*. He said *Veratrum* and *Camphor* and other medicines were given by the previous physicians to no effect. This time *Ricinus* was so efficacious, and that stu-

4

dent (studying in the old school) gave the authority of Dr. George Johnson as the promulgator of *Castor oil* treatment in cholera. Indications: Purging and vomiting of rice-water fluid; cramps in the extremities; there is seldom or no pains in the abdomen; extreme prostration; complete suppression of urine; scarcely perceptible pulse; very slight coldness of the extremities.

In *Ricinus* cases there is gradual sinking of the vital power: in this respect it differs both from *Camphor* and *Veratrum* which have rapid sinking. The stools may be sometimes tinged with bile, or a little slimy and mixed with mucus and blood.

I generally use the 6th decimal dilution after each stool.

Jatropha cur. produces depression of heart. Vomiting is more prominent than purging. Indications are whitish vomiting, like white of egg; stools watery and in gushes, gurgling and rumbling in the bowels, cramps in calves, pains and burning in stomach. There is coldness with slight perspiration and thready pulse. "Watery diarrhœa as it spurted from him." The alarming symptoms of cholera are not marked in this remedy, and the patient is devoid of any anxiety for his future. In fact he cares nothing for his disease and is cheerful.

Euphorbia is another remedy closely analogous in its action to *Jatropha*. Both these remedies, in fact, are medicines for choleraic diarrhœa rather than true Asiatic cholera. There is very little difference in their action. I use the 6th decimal dilution, a dose repeated after each evacuation.

Croton tig.—Though it is not a medicine for true cholera, it often cuts short the disease by its early administration, otherwise it would be developed into a terrible disease; stools are yellow, watery, passed forcibly like shot; worse after food and drink; deadly nausea; vomiting after drinking and great prostration. Sixth decimal or 30th centesimal may be used.

Antimonium tart.—It is very closely related *Veratrum alb.*, and I often use it when that remedy fails to act.

Indications: Very much like *Veratrum*. Purging of rice-water stool; vomiting with great effort; cold, clammy perspiration; drowsiness with complete exhaustion; almost imperceptible pulse; heart's action failing; labored and difficult respiration; and complete collapse. When cholera breaks out during an epidemic of smallpox, it is better to make a choice of *Antim. tart.* in the very onset of the disease. We have several times witnessed the charming effect of the remedy in such cases.

Iris versicolor is very efficacious in checking cholera evacuations, but it is a remedy for what is called English cholera. I used it in cases where vomiting is a prominent and distressing symptom. Acidity, with burning sensation in the whole alimentary canal and bilious vomiting. In the case of a young gentleman in a suburban town of Calcutta, I got a charming effect from *Iris*. The attending physician tried all medicines to check purging and vomiting, without any effect. *Iris versicolor* was given by me, and the patient was all right within a couple of hours. He had been suffering for two days before my arrival. These are the principal remedies in the developed stage of cholera. They are more or less potent in checking cholera evacuations.

But if the evacuations are not stopped, or stopped after doing considerable

damage to the system, the case goes on
to the next stage — the *collapse*. In
this stage the patient is on the point of
death; in fact, all the signs of death
are visible on him. We must not lose
heart on seeing these serious symptoms, as we can still do a good deal of
good to save the patient. The following remedies are to be thought of in
the collapse stage:

Arsenicum album. {
Aconite.
Camphor.
Veratrum alb.
Carbo vegetabilis.
Cuprum arsen.
Hydrocianic acid.
Cobra.
Secale cor.
Antim. tart.
}

Practically a great deal of difficulty
would arise in treating this stage of
cholera. We have so many remedies
closely analogous to each other, that it
seems almost impossible to select one.

If we take pains to record the symptoms very minutely, our difficulty would be much minimized and we may come to a definite and reliable selection.

Arsenicum album. is a very important medicine in the collapse stage of cholera. Its pathogenetic symptoms are so closely similar to Asiatic cholera that an arsenical poisoning case may be easily mistaken for a cholera case. It has a vast range of action, and we have repeatedly verified its curative power in most serious cases of the disease. It is for this reason that I select it as a prototype of collapse remedies. Its symptoms are very marked and unmistakable.

Great irritability associated with profound exhaustion is the prominent characteristic of *Arsenic* cases. You will see patients whose pulse vanishes, great weakness, even unable to utter a single word, yet restless, anxious and irritable.

Indications for *Arsenic*: Great anxiety and restlessness; fear of death; great prostration of strength; sunken eyes, distorted face, pointed nose; cold and clammy perspiration; burning of the whole body; retching and vomiting; unquenchable thirst, drinking frequently but small quantities at a time, and vomiting immediately after drinking; violent burning of stomach and abdomen; urine completely suppressed, thin, watery stools.

When a patient gets an attack of cholera after eating too much fruit and drinking iced water, living in a damp place, exposed to the influence of putrefactive and offensive smells, *Arsenic* is the remedy.

I often use the 30th dilution in frequently repeated doses until favorable symptoms are observed. I have many a time saved very desperate cases of cholera by giving frequent doses of the 200th dilution when the 30th failed.

Arsenic has been frequently administered indiscriminately, without reference to its symptomatic indications. This is indeed a bad practice. It is, therefore, as Dr. Bell says, that *Arsenic* does more harm than good in the hands of ignorant persons.

Camphor has been used in the stage of collapse, but I have not found it very efficacious; so generally doubt resort to it in collapse. Indications: sudden and rapid prostration; coldness of surface; cold sweat; bluish countenance; husky voice; violent cramps and loss of consciousness. It should be cautiously given, and as soon as signs of improvement begin, or warmth returns, it must be stopped.

Aconite is pre-eminently the best remedy in the collapse stage of cholera. Dr. Richard Hughes says that in time *Aconite* will be a valuable medicine in cholera. Dr. Hempel is the first physi-

5

cian who draws our attention to its use in cholera. Dr. Hughes wrote this long ago, and I believe the time is now come, and have used *Aconite* very extensively and with good results.

Indications: Great anxiety and fear of death; great coldness of the whole body; cold perspiration; great thirst; restlessness; labored breathing, with pains and oppression of chest; hardly perceptible, or thready and quick pulse; weak and slow beating of the heart. In cases of violent colicky pains in abdomen, it is one of our greatest helps. An elderly lady had an attack of cholera, collapse, restlessness and unbearable pains in the epigastric region. Many Homœopathic remedies were tried without effect. I found her in great agony, gave her *Aconite* 1x every half hour, and in two doses she was relieved of her pains and reaction took place. In warm days with cool nights, and after exposure, I find *Aconite* very efficacious.

Carbo vegetabilis is one of our most reliable remedies in collapse of cholera. I employed it in very many cases where death seemed inevitable, and with good results. When the reactive power of the system is gone, *Carbo veg.* is indicated. Indications: The patient lies as if dead; no signs of irritability about him; pulseless; cold and clammy sweat; leaden hue of the body; husky voice; difficult and labored respiration; no thirst; no purging and vomiting; abdomen often distended; difficulty of breathing and suppressed urine. Lower dilutions are of no use; 30th and upwards are to be employed.

Hydrocyanic acid.—It is a marvelous remedy, acting promptly, and sometimes snatching away the patient from the verge of death. On one occasion I had to attend a little girl. When I arrived I saw her gasping her last breath. She could not swallow medicine or anything else. I poured a few drops of *Hydrocyanic acid* on a clean handkerchief and put it in her nostril, and to my utter surprise I found her breathing quietly in a few minutes; and she made a perfect recovery. It is for this reason that our esteemed colleague, Dr. Mohendra Lal Sircar, speaks of it as follows: "If any remedy is entitled to be spoken of as a charm, it is *Hydrocyanic acid*. It would seem at times to restore animation to a corpse." Indications: Icy coldness of the body with pulselessness; breathing slow, deep and somewhat spasmodic in character, beating of the heart slow and weak; urine and stools generally suppressed.

Third or sixth decimal dilutions are generally used more frequently almost every half hour or so. It should be freshly prepared. Cyanides are often useful in trituration.

Cobra or Naja trip.—This is a medicine from the poison of a most venomous snake of India. The effect of this poison is very swift to our people. Compare it to a sudden attack of Asiatic cholera. Dr. Salzer, in his excellent book on cholera, speaks of it and other snake poisons in the following words: "We administer them when respiration quickens, becoming at the same time more and more superficial, while the heart's action is normal and still comparatively vigorous. This sort of respiration is a sure sign of impending paralysis of the respiratory centre, and coincides in so far exactly with what occurs under the venomous influence of snake poison." Higher dilutions are better.

Secale cornutum.—It is another very important medicine in the collapse stage of cholera. Indications: Watery, slimy and offensive stools; vomiting of water; eyes sunken; violent cramps of the calves, the hands and the chest; great restlessness; constant thirst; difficulty of breathing; pulselessness, or small, slow and almost imperceptible pulse; coldness of the body, but patient feels heat inside, and cannot keep clothes on body. I have very little confidence in *Secale* in collapse, but it is a very valuable remedy in some other conditions in cholera. It often removes distressing cramps when *Cuprum* fails. I frequently find it efficacious in removing that dreadful symptom, the cramps and pains in the side of the chest, more so in heart. Appearance of menstrual flow during an attack of cholera is very serious, and in such cases *Secale* proves of immense value. Typhoid condition in cholera is also a very dangerous complication, and here *Secale* is our great help.

Febrile heat after coldness; sleepiness with now and then restless and often profound comatose sleep; pinched appearance; frequent and small pulse. It may be useful in gangrene and bedsore, ulceration of cornea and some other symptoms, derived from the low vitality of the system after an attack of cholera. Dilutions from 6th to 30th best.

Veratrum alb. is also a very useful remedy in collapse. Dr. Salzer writes as follows: "I can hardly believe that *Veratrum* should not be as useful in collapse owing to the paralytic condition of the heart. Perhaps we give the remedy at too long intervals. Dr. Carrol Dunham recommends it to be given like *Camphor*, every five minutes. Much of this disrepute of the drug in collapse may also be owing to its not having always been administered at the right place and according to right indications.

Antimon. tart. being a depressant

medicine on the heart, is recommended in the collapse stage of cholera, and we often find it useful. My indications are generally the same as *Veratrum*. I find it very useful in cholera with drowsiness and complete exhaustion.

Another medicine of collapse is *Nicotine*, the active principle of tobacco. It may be used in collapse with cold sweat, deadly nausea, sleepiness and weakness of heart's action. It is sometimes applicable in typhoid symptoms with vomiting and drowsiness. I have had very little experience with this remedy.

COMPLICATIONS AND SEQUELÆ.

Our troubles do not end with the successful treatment of the collapse stage; various other ailments await our poor patient, and we must be on our guard to combat them in proper time. These must be considered as serious as a developed or collapsed stage of the disease. Many of our patients often succumb after reaction from these complications, etc.

Uræmia.—If urine is not voided after reaction fully sets in, we must do something for it. Some physicians are in a great hurry about establishing a urinary secretion; they become so impatient that they want it even in a collapsed state. When reaction is in its progress, we must stop all medicines and wait. If still urine is not passed, and there is fear of a typhoid state supervening, then we stir up and do something. Many a time our previously employed remedies are sufficient to restore urinary secretion; so, without abandoning them for new remedies, we must continue them according to the symptomatic manifestation of the case. *Arsenic, Hydrocyanic acid* and *Tabacum* or *Nicotine*, have to be prescribed.

But if they fail, and if there is impending congestion of the brain and its

6

symptoms, we must resort to *Bella-donna*. I generally find it useful in the 30th dilution. *Hyosciamus* 30th or sometimes 200th is often very benefi-cial with the nervous and other typhoid symptoms. *Opium* is one of our best remedies if there is comatose condition along with uræmia.

We have seen *Agaricus muscarius* and *Muscarin* valuable in the typhoid condition from uræmia associated with pulselessness or small, thready pulse, coldness of the surface and muttering delirium. *Muscarin* may be tried in the early stage of cholera in purging and vomiting.

If urine is collected in the bladder and there is sufficient expulsive power, and patients exert their utmost but void no urine, I use *Cantharis* 6x or 30th. Failing with this remedy, and if there be some burning in the ureth-ral canal and no urging, *Terebinthina* may be administered. Our much-vaunted *Kali bich.* produced no effect in my hands. I am in the habit of using *Carbolic acid* in cases of uræmic intoxication and delirium in cholera cases with the following symptoms: constantly agitated, with a piercing cry, delirious starting from sleep; tongue dry, coated with thick yellow fur; great thirst and a high fever; urine may be dark, black or olive green in color. It is sometimes suppressed.

In the reaction stage we sometimes meet with feverishness, and when it is slight we must not give any medicine; it generally goes off itself. When it is assuming a grave aspect, we may try *Aconite* and *Veratrum*, all according to indications. *Belladonna* has the power of checking it when above remedies fail. *Rhus tox.* and *Bryonia* may be indicated. *Phosphoric acid* is one of our best remedies in mild forms of fever from weakness of the system. This later remedy is also useful when reactionary fever assumes a typhoid character.

Hiccough is often a very distressing symptom. Our ordinary hiccough remedies are not very efficacious in this state. *Cuprum met.* and *Cuprum arsen.* are my great helps. *Veratrum, Nicotine* and *Hydrocyanic acid* are also recommended. If there is some faulty condition of the digestive system we may think of *Nux vomica, Cicuta v., Phosphorus* and *Ignatia. Belladonna* may be useful. Ice water sometimes has a very prompt action in checking hiccough.

We have sometimes seen patients, cured to all intents and purposes, die suddenly of dyspnœa. Dr. Macnamara is of opinion that in these cases there is formation of a clot in the right side of the heart, usually extending into the pulmonary arteries. Dr. Salzer, on the authority of Dr. Buchner, advised us to give *Calcarea arsen.* 6th or 12th in such cases. But death is so sudden in these cases that nothing can be done.

After the cholera symptoms are over we often get some cases of obstinate diarrhœa as the after effect of the disease. In these cases where the stools are yellow, watery, quite copious and sudden, *Croton* gives prompt relief; when there is colic with saffron-yellow, watery stools and much prostration, *Colchicum;* when stools are yellowish or white, painless, *Phosphoric acid* or *Podophyllum;* in tympanitic distention, with passage of flatus, stools yellow, *Natrum sulph.;* in retarded convalescence, with diarrhœa of undigested food, *China.* A dose or two of *Sulphur* high is needed to complete the cure. When there is a tendency towards dysentery, stools greenish, with colic, *Merc. sol.,* but if bloody and slimy, *Merc. cor.* may be employed.

Vomiting often becomes troublesome and persistent, and defies all our well-selected remedies. In such cases some bland and mucilaginous diet is all that is necessary to check it. I generally give rice water, arrowroot or barley water, with a little salt, and acidulated. Other complications should be treated as general diseases.

Ulceration of the cornea requires *China*, *Mercurius*, *Secale* and *Argentum nitricum*.

Diet.—There is a good deal of discussion among the physicians on this subject. From brandy, stimulants, broth, milk and others, to no diet, have been prescribed by medical men. From practical observations in many cases we are of opinion that during the progressive and collapse stages of the disease no food should be given except ice and plenty of cold water to appease thirst and cool down stomach. After the tempest is over we can give a little barley or arrowroot water. When after a time craving for food occurs, and a gradual addition of nutritious but easily digestible food may be tried. The stomach becomes very sensitive after an attack of cholera, so particular care is necessary in selecting food. I have seen fatal cases of relapse from improper food.

THERAPEUTICS OF CHOLERA.

ACONITUM NAPELLUS.

Aconite is a great anti-cholera medicine, but it should be used with proper indications. Dr. C. J. Hempel is the first physician who strongly advised us to use this remedy in cases of cholera.

Characteristics.—*Aconite* is most frequently indicated in most recent cases of illness occurring in young persons, especially in girls of a full, plethoric habit, and who lead a sedentary life; persons who are easily and readily affected by sudden atmospheric changes.

On rising from a recumbent position the red face becomes deathly pale, or he becomes faint or giddy and inclined to fall over; in consequence he fears to rise again. These symptoms are often accompanied by vanishing of sight and unconsciousness.

Great fear and anxiety with consequent nervous excitability; afraid to go out, to cross the street or to go into a crowd where there is any excitement or many people. The countenance is expressive of constant fear; life is rendered miserable by it. For the effects of mental shock. Increased sensibility; the whole body is sensitive to touch. Many of the symptoms are accompanied by shivering.

Hahnemann said: "*Aconite* should not be given in any case which does not present a similar group of symptoms." These are the symptoms of the mind and disposition, viz: Restlessness, anxiety and uneasiness of mind and body, causing tossing and sighing and frequent changes of posture; foreboding anticipations of evil, anguish of mind, dread of death, and even distinct antic-

7

ipation of its occurrence. (Dunham.)

In cholera these mental symptoms are of very frequent occurrence, and *Aconite*, therefore, is one of our greatest helps.

It happens that *Aconite* is frequently indicated at the very beginning of some acute affections, and that if properly used in such cases it will often cut short the career of the disease. From these facts has arisen a fashion of giving *Aconite* almost as a routine prescription in the beginning of all acute cases indiscriminately, particularly if the cases are supposed to be characterized by that protean phantom of the pathologists — inflammation. Great mischief often results from this practice negatively, inasmuch as it causes the loss of valuable time during which the true specific remedy which should have been given at the very first might have been acting; and positively, inasmuch as the *Aconite* often, when improperly administered, does real mischief, exhausting the nervous power of the patient, and adding to the nervous prostration which is already probably a great source of danger. *Aconite* should never be given to save time. It were better to give nothing, because *Aconite*, if given in a case which does not call for it, might do mischief.

Aggravation.—In the evening; night, especially after midnight; in a warm room; from taking cold; drinking cold water; from tobacco smoke; lying on either side; on coming from the open air into a warm room.

Amelioration.—In open air; while at rest (except at night in bed); from perspiration; from urine; lying on the back.

Cause.—When disease takes place from dry, cold west or northwest winds; during hot days with cool nights in summer; made worse by getting wet, especially getting the feet wet; from suppressed perspiration by uncovering or sitting in a draught; rheumatic exposure of any kind; by fright; after eating fruit.

Stools.—Watery; green or black; dysenteric, bloody and slimy; frequent but small; involuntary; pains in abdomen.

Sleeplessness; frequent dry heat; full, hard, very quick pulse.

In cholera, hippocratic countenance; face bluish; lips black; expression of terror; cold limbs with blue nails; collapse. (Bell.)

ANTIMONIUM TARTARICUM.

It is a very useful remedy in cases of cholera in its first, and in the last stage. It is a close analogue of *Veratrum album* and *Arsenicum* in the treatment of cholera.

Characteristics.— Great prostration; cold sweat; thready pulse; continuous nausea; *straining to vomit, with perspiration on the forehead.* Great irritability. Vomiting of greenish, watery, frothy mucus or food. Vomiting is accompanied by trembling of hands and fainting, and is followed by great languor, drowsiness, loathing, desire for cooling things; pale, sunken face; dim and swimming eyes.

Violent and painful urging to urinate, with scanty and bloody urine.

Drowsiness or somnolency.

Palpitation of the heart.

Stools.—Watery; thin, bilious, liquid and greenish; brownish, yellow, fæcal; profuse; of cadaverous smell; colic in abdomen.

Aggravation.—During exanthemata; after taking cold in summer; at night.

ARSENICUM ALBUM.

It is one of the most important and deep-acting remedies in cholera. But it requires a great deal of caution to use it, for in ignorant and careless hands it produces much harm. It is for this reason, Dr. Bell says: "There is reason that as routine is easier than study, *Arsenicum* may have accomplished more harm than good in the hands of Homœopathic practitioners. No remedy has been more frequently given in acute affections of the bowels, while it is not the most frequently indicated, and it is not a remedy to be unwisely used."

Characteristics. — Great prostration; rapid sinking of the life forces. Depressed, melancholic, despairing, indifferent, fearful, restless, anxious and full of anguish.

Great restlessness, constantly changing place.

Violent, unquenchable thirst, with frequent drinking of small quantities of water. Desire for acid, cold water and spirits.

Vomiting after eating or drinking. Violent pains in the stomach and burning in the stomach and abdomen.

Urine retained or suppressed.

Stupor with hot skin; twitching of limbs and tonic spasms of the fingers and toes.

Skin cold and covered with clammy perspiration, though the patient complains of intense burning heat internally.

Great weakness, fainting and rapid exhaustion. Very rapid or scarcely perceptible pulse.

Restless sleep, broken by starts and convulsions.

Dunham says: "The fact cannot be too often called to mind, nor too strongly insisted upon, that our most characteristic indication for the use of a drug which presents such well-defined general symptoms as *Arsenic* does, and indeed as every other well-proven drug does, are derived not from its local action upon any organ or system, not from a knowledge of the particular tissues it affects, and how it affects them, but upon general constitutional symptoms and their conditions and concomitants. If this were not so, in the presence of how many maladies, of the intimate nature of which we are wholly ignorant, and which nevertheless we cure, should we be utterly powerless for good?"

In diarrhœa after eating and drinking, *Arsenic* is a good medicine.

Stools.—Dark color, watery; offensive odor; greenish, watery and mucous; often painless. Burning of the rectum and anus.

Aggravation.—After eating; at night; after midnight (1 to 3 a. m.); from cold food and drink.

Amelioration.—From external heat (pains) and cold air.

CAMPHOR.

Camphor is one of our very important medicines in cholera, especially in the very beginning of the disease. It may be also useful in the collapse stage. Dunham says: "In *Camphor* collapse is more prominent; in *Verat. alb.*, the evacuations and vomiting; in *Cuprum*, the cramps."

Characteristics.—Sudden and great sinking of strength; great anguish; coldness of the whole body; weak voice; violent or no thirst; wild, staring and unconscious look.

The vomiting and diarrhœa suddenly cease, and the child becomes unconscious.

8

In cholera, great sinking and collapse, sometimes without stool or vomiting. Cold as death, but cannot bear to be covered. (Bell.)

Stools.—Blackish; dark brown; large, thin and involuntary; like rice water; generally painless; attack sudden.

Aggravation.—During cholera epidemic; in summer; in pernicious fever.

CARBO VEGEBILIS.

It is one of our prime remedies in cases of cholera in its collapse stage, when all reactive power of the system is at abeyance. Many physicians and other persons laugh at the medicinal properties of charcoal. They say it is an inert substance, but Hahnemann proved and showed its clinical virtue. He says: "Their (carbons generally) medicinal virtues are latent, and can only be elicited by trituration and succussion as taught in Vol. I. 'Chronic

Diseases.'" Dr. H. C. Allen very justly remarks: "It is only after extended potentization that the symptoms on healthy persons steadily increased, so that to the proving of the triturated (potentized) vegetable charcoal we are indebted for the greatest discovery in the history of medicine, viz: *that the dynamic is the only curative force in drugs.* This great discovery of Hahnemann is certainly destined to revolutionize the art of healing."

Characteristics.—Symptoms of imperfect oxidation of blood. For the complaints of persons, either young or old, who have suffered from exhausting diseases—diarrhœa, hæmorrhages, profuse sweat and suppuration.

Weak digestion: the simplest food disagrees. Excessive accumulation of gas in the stomach or intestines, temporary relief from eructation. Want of susceptibility to medicinal action.

Want of nervous irritability or of the vital reactive power.

Hectic fever; exhaustive night sweat, profuse, putrid, sour perspiration ; sallow complexion, sunken features, vital forces nearly exhausted.   (Hering.)

Characteristics.—In cholera: collapse without stool.   Nose, cheek and fingers icy cold. *Desire to be fanned.* Hiccough and vomiting.   Greenish, pale color of the face.   Sopor without vomiting, stool or cramps.   Congestion of blood to the head and chest.   Pulse thready, intermittent and scarcely perceptible.

Stools.—Bloody, brown, watery and slimy ; black and semi-liquid ; frequent, involuntary.

Aggravation. — From chilling the stomach with ice water when overheated.   After long-continued severe acute disease.   After loss of fluid.

CUPRUM.

Cuprum is one of our most trusted and efficacious remedies in cholera; always useful from the beginning of an attack to its collapse stage. It is especially valuable in severe cramps of the extremities. Both *Cuprum metallicum* and *aceticum* are used.

Hahnemann was a great advocate of *Cuprum* in cases of cholera.

Characteristics. — Restless tossing about and constant uneasiness. Face distorted and blue. Sunken eyes. Excessive thirst; drink descends the œsophagus with a gurgling sound. Deathly nausea and retching; vomiting of water with flakes, with violent colic and cramps. Violent pains in the stomach. Feeling of constriction beneath the sternum. Violent spasm in the abdomen and upper and lower extremities, with piercing screams. Dyspnœa so great that the patient cannot bear a handkerchief before the face. Sighing respiration. Urine suppressed or scanty. Comatose sleep after vomiting. Intense coldness and blueness of surface, with cold sweat and great prostration.

Uræmic convulsion with loquacious delirium, and followed by apathy and collapse.

Aggravation.—During epidemic cholera.

Amelioration.—From drinking cold water (vomiting).

Stools.—Watery, with flakes. Frequent but not quite copious.

HYDROCYANIC ACID.

It is our sheet anchor in severe cases of cholera collapse; in fact, it often brings a corpse to life. Dr. Hering says: "In the last stage of Asiatic cholera, when diarrhœa has ceased and vomiting decreased, when there is anguish, with pressure on the chest, and the patient becomes cold, with gradual extinction of pulse.

Characteristics.—Despondency and oppression. Vexed mood. Loss of consciousness. Eyes distorted and half open. In Asiatic cholera there is sudden and great prostration. Absence of thirst or violent thirst; fluid runs down the œsophagus with gurgling noise. Epileptiform attacks of fainting. Vomiting of a black fluid. Intense gastric pains.

Retention of urine. Uræmia and asphyxia from uræmia.

Noisy, agitated and spasmodic breathing. Tightness of chest and sense of suffocation. Rattling, moaning and slow breathing. Dyspnœa. Irregular and feeble breathing of heart, pulse hardly to be felt. General weakness, loss of power and great exhaustion.

Stools.—Involuntary stools, hiccough and great prostration. Sudden cessation of discharges. Rapid progress towards asphyxia. Green, watery stool.

Aggravation.—After eating and drinking. In the evening. After cold food.

IPECACUANHA.

This medicine is especially suited to cholera infantum and gastric variety of cholera. In cholera infantum it may be followed by *Arsenic.* (Bell.)

Characteristics.—Dejected and morose. Irritability and impatience. Cold sweat on forehead. Eyes sunken and blue margin around them. Thirst or thirstlessness. Hiccough with nausea. Nausea distressing; constant; empty eructation; accumulation of saliva in mouth; qualmishness; vomiting of ingesta, bile, grass-green mucus, putrid, blood or pitch-like substances or sour fluid.

Stools.—Green mucus, as green as grass; greenish, watery. Bloody and fermentable. Bilious. Dark, almost black, like frothy molasses. Putrid and frequent.

Aggravation.—In the evening or night. During dentition. After unripe fruit and vegetables.

Amelioration.—From rest.

IRIS VERSICOLOR.

It is a useful medicine in what is called English cholera or summer diarrhœa or cholera morbus. It may be used in Asiatic cholera with rice-water stools, cramp and other symptoms, but not so frequently.

Characteristics.—Low spirited. Fear of approaching illness. Sunken eyes.

Acid and sour eructation. Heartburn. Nausea and vomiting of sour fluid, food or sweetish water, soured milk in children, yellow or viciated

9

bile. Burning from mouth and anus. Violent efforts to vomit. Vomiting of extremely acid fluid, which excoriates the throat.

Stools.—Watery, watery with mucus. Premonitory diarrhœa with vomiting and cramps. Involuntary evacuation of rice-water character. Body icy cold.

Aggravation.——At night, from 2 to 3 a. m. In hot weather.

Amelioration.—Of colic by passing flatus and bending double.

PHOSPHORIC ACID.

This is a medicine for premonitory diarrhœa and cholera morbus, often cutting short the disease. It is also useful in cases of mild delirium and in convalescence when general weakness of the system is the predominant character.

Characteristics.—Apathy and indifference. Somnolency. Quiet delirium.

Pale and waxy complexion. Bleeding from gums. Thirst for large quanties of water. Distention of abdomen with gurgling noise. Profuse emission of pale, watery urine which debilitates the patient; opaque, milky white sediment in urine.

Profuse debilitating perspiration at night. Cramps in the extremities, especially the upper.

The patient gains flesh in spite of diarrhœa.

Stools.—White watery; thin yellow with sediment; whitish, gray, fæcal; undigested; involuntary; painless and very offensive.

Aggravation.—From depressing mental emotion; young persons who have grown rapidly. After stool.

RICINUS.

The use of this remedy is comparatively new in the Homœopathic school for cases of cholera. Drs. Salzer and Bhaduri are among the foremost I know who gave it a fair trial in cholera cases in India, but Dr. Hale, of Chicago, gave us a hint long before for its use in cholera cases.

Characteristics.—Indifference; no fear or apprehension. Profound exhaustion. Face pale and death-like. The skin cold and shrunken. Drowsy. Loss of consciousness. Vertigo. Buzzing in ear. Collapse. Burning thirst. Nausea and vomiting. Vomiting of rice-water substance or bile. Cramps in the stomach. Complete suppression of urine. Attack slow and insidious, often from indigestion and diarrhœa.

Stools.—Rice water. Bilious watery. Mucous and bilious. Fæcal, bloody and slimy.

Aggravation. — From eating and drinking. In summer.

SULPHUR.

It is a remedy in almost all diseases, and so it is useful in cholera. Many Homœopathic physicians have an impression that *Sulphur* is only used in chronic cases, so it is not applicable in such an acute disease as cholera, but they are mistaken. I have administered it in various stages of cholera with most gratifying results. Of course it must be given with a due care as to its proper indications.

Sulphur may be said to be the central remedy of our Materia Medica. It has well-defined rotations with every drug we use. The great utility of *Sulphur* arises from this peculiarity; it is our mainstay in defective reaction. When the system refuses to respond to the well-selected remedy, it matters not what the disease may be, whether it corresponds characteristically with the syptomatology of *Sulphur* or not, it will often be the remedy to clear up the case, or pave the way for another drug which will cure. (Farrington.)

Characteristics.—Anxious and fearful. Melancholy; depressed about illness.

Eyes sunken, surrounded by blue margin. Face pale and collapse with expression of anxiety. Tongue brown or coated white.

Thirst much and drinks large quantity. Desire for sweets but aversion to meat. Hiccough; eructation empty or tasting of food. Hunger and voracious appetite, with burning of stomach. Griping or pinching colic in abdomen. Distension and hardness of abdomen. Emission of fetid flatus. Rumbling and gurgling in bowels.

Retention or total suppression of urine. Frequent urging to urinate. Only a few drops discharged, with burning in the urethra. Feels suffocated; labored and heavy breathing.

Pulse feeble, full, hard, accelerated, at times intermittent. Hands and feet icy cold or burning. Cramps in the leg. Sleepiness in the evening, night restless. Sleeping with half-open eyes. Screaming out in sleep. Jerks and twitching during sleep. Patient lies in dull and stupid state, with muttering delirium, talking incoherently. Skin dry and flabby. Offensive odor of body, despite frequent washings. Flushings of heat. Chilliness about the lower part of body. Excessive prostration and rapid emaciation.

Stools.—Sudden call for stool on waking in morning. Stools watery, brown or green. Watery, thin fecal, changeable, hot, frothy, sour smelling or fetid. Involuntary, with sudden explosion.

Cholera infantum generally begins after midnight; diarrhœa and vomiting; discharge from bowels watery; green and involuntary. Sometimes sour smell, at other times offensive. Vomiting frequent, often sour, with cold perspiration on face. Face pale, fontanelles open, hands and feet cold. Child lies in stupor with eyes half open, not much thirst and entire suppression of urine; psoric patients prone to eruptions and excoriations; hydrocephaloid symptoms.

Cholera Asiatica: As a prophylactic, a pinch of powdered milk of *Sulphur* worn in stocking, in contact with soles of feet; diarrhœa commences between midnight and morning with or without pain, with or without vomiting, ineffectual desire to evacuate; diarrhœa and vomiting at same time; numbness of limbs, cramps in soles of feet and coldness; blueness under eyes; coldness of skin; indifference of mind; during convalescence, red spots, furuncles, etc; susceptibility to temperature, warm things; feet hot; nerve symptoms.

Aggravation.—Early in the morning in bed. In the evening and after midnight. After taking milk. After suppression of eruptions. During sleep.

Amelioration.—By heat and dry heat on abdomen (colic).

VERATRUM ALBUM.

The use of this remedy in cholera is very extensive and effective. I have been frequently in the habit of using it rather in the fully developed stage of the disease, and with good results. *Veratrum* is a remedy of great value, and one often required, but like all others it demands a careful selection, and is not to be given in every case of cholera morbus or cholera. The most characteristic symptoms are the same in both cases, only more violent in the latter. The immediate accompaniments of the stools, with the thirst and cravings, distinguish this remedy. *Veratrum* is seldom indicated in painless cases.

Characteristics.—Depression of mind and despondency. Fear and anxiety.

Cold sweat on the forehead, with anguish and fear of death.

Face: collapsed; pale, bluish; nose pointed, of leaden hue, alternately pale and red, sunken, with anxious expression; pinched up, death-like, hippocratic. Eyes sunken, distorted and turned up. Grinding of teeth. Tongue coated yellowish brown, back part black.

Thirst great and, during perspiration, for large quantities of very cold water and acid drinks; craves fruits and juicy things.

Hiccough. Vomiting forcible, excessive, violent, with continual nausea, retching and great prostration. Vomiting of thin, yellowish substance and of rice-water character. Pains in stomach as from ravenous hunger. Sinking and empty feeling in abdomen. Violent colic. Abdomen distended and sensitive.

Retention of urine, frequent and continuous urging to urinate.

Voice weak and hoarse.

Respiration weak, cold breath, constriction of chest. Typhoid fever, especially during cholera season, when vital forces suddenly sink. Coma.

Cholera Infantum: Attack sudden; violent watery purging and vomiting; cold surface; prostration; vomiting, excited by least amount of liquid taken; great thirst for large quantities of water; excessive weakness, stools watery and inodorous; tongue and breath cold; difficulty of breathing; desire to sit up; blueness around the eyes; skin becomes drawn tightly over bones of face; wrinkling of skin of hands and fingers; sensation over abdomen.

Cholera Morbus: Increased at night; cold sweat on forehead; vomiting and purging at the same time; after fruits; with profuse brownish discharges; thirst; cramps in the calves, feet and fingers; prostration; great weakness after stools.

Cholera Asiatica: Great torpor of vegetative system, without any great mental or sensory disturbance; little depression of spirits; fear of death or indifference; vertigo; violent evacuations upward and downward; icy coldness of the body; great debility; cramps in calves; vomiting, with constant desire for cold drinks; copious, watery, inodorous stools mixed with white flakes; face pale, without any color or bluish, blue margins on the eyes; deathly anguish in features; cold tongue and breath; hoarse, feeble voice; great oppression and anguish in chest, giving patient a desire to escape from bed; violent colic, especially around the umbilicus, as if the abdomen would be torn open; abdomen sensitive to contact, with drawing and cramps in fingers; wrinkled skin of palms of hands; retention of urine.

Stools.—Watery, with flakes; rice-water, watery and greenish, frequent, profuse and sometimes painless.

Aggravation.—In summer; by drinking; after fruits and from taking cold.

REPERTORY.

Cholera: *Acon., Ars., Camph., Carbo veg., Cicuta, Colch., Cupr., Hydro. acid, Jatr., Merc. sol., Phos., Phos. ac., Podo., Scc., Sulph., Tabac., Verat.*

asphyctica s. sicca: *Camph., Carbo veg., Laur., Tabac.*

infantum: *Acon., Apis mel., Æthu., Ant. c., Ant. t., Arg. nit., Ars., Bell., Bis., Calc. c., Camph., Carbo v., Colch., Coloc., Crot. t., Elat., Grat., Helleborus, Ipec., Iris v., Jatr., Kali bich., Laur., Phos., Podo., Sarsap., Sec., Sulph., Tabac., Verat., Zincum.*

morbus: *Acon., Ant. c., Ant. t., Ars., Camph., Colch., Coloc., Crot. t., Elat., Grat., Ipec., Iris v., Jatr., Kali bich., Phos., Phos. ac., Podo., Ricinus, Scc., Tabac., Verat.*

CHARACTER OF THE STOOLS.

Albuminous: *Diosc., Jatropha, Natr. m.*
Attack, sudden: *Camph., Cupr., Scc.*

Stools, bilious: *Acon., Æthu., Agar., Aloe, Ant. tart., Ars., Carbol. ac., Cham., China, Cina, Coloc., Corn. c., Cub., Diosc., Dulc., Elat., Fluor. ac., Ipec., Lept., Lil. tig., Merc. c., Merc. v., Phos., Podo., Puls., Sulph., Verat., Zinc.*

brown: *Ars., Camph., Merc. c., Sulph., Tart , Verat.*

greenish: *Ars., Bell., Canth., Cham., Ipec., Laur., Merc., Nux vom., Phos., Phos. ac., Sulph., Verat.*

gray, or slightly whitish: *Acon., Ars., Bell., Carbo v., Cham., Lach., Merc., Phos , Phos. ac., Puls., Rhus, Sulph., Verat.*

liquid: *Arn., Ars., Carbo v., Chin., Cic., Jat., Lach., Meph., Phos., Phos. ac., Secale, Verat.*

liquid and whitish: *Ars., Camph., Cupr., Jat., Phos., Phos. ac., Sec., Verat.*

liquid and whitish, with white coated tongue: *Cupr., Phos., Sec.*

liquid, with continual pain at the pit of the stomach: *Ars., Camph., Chin., Cupr., Phos., Verat.*

liquid, with rumblings in the intestines: *Ars., Jat., Nux v., Petr., Phos., Phos. ac., Puls., Rhus, Sec., Sulph., Tart., Verat.*

Stools, liquid, evacuation painful (attended with colic): *Ars., Carbo v., Phos., Spig., Staph., Verat.*

evacuation painless: *Ars., Carbo v., Chin., Cic.. Phos , Phos. ac., Sec., Spig., Verat.*

mucous and watery: *Ars., Bell., Chin., Ipec., Nux v., Phos., Phos. ac., Puls., Rhus, Sec., Sulph. ac., Tart.. Verat.*

like rice water, or stools like whey or water, with whitish or grayish flocks in it: *Ars., Camph., Cupr., Ipec., Jat., Phos., Phos. ac., Secale, Verat.* If there is inflammation, consult also *Acon., Bry.* and *Rhus t.*

like rice water, or watery, grayish, whitish and flocculent, with great thirst: *Ars., Camph., Cupr., Ipec., Phos., Phos. ac., Verat.* With fever, *Acon., Bry.*, or *Rhus tox.*

Acon., Ars., Camph., Bell., Cham. Chin., Cupr., Ipec., Jat , Merc., Nux v., Phos., Phos. ac., Sec., Sulph., Verat.

watery and white, flocky, with cramp and thirst: *Acon., Ars., Bry., Camph , Cupr., Ipec., Phos., Phos. ac.. Rhus tox., Sec., Verat.*

II

Stools watery and white flocky, with clonic spasms (spasmodic movements) and thirst: *Acon.*, *Ars.*, *Bry.*, *Camph.*, *Cupr.*, *Ipec.*, *Phos.*, *Phos. ac.*, *Sec.*, *Verat.*

whitish flocks, with pulselessness or scarcely perceptible pulse: *Acon.*, *Ars.*, *Bry.*, *Camph.*, *Phos. ac.*, *Rhus*, *Sec.*, *Verat.*

watery or liquid, with white flocky in grains, having the consistence and color of tallow: *Phos*

yellow (especially in the early stage of the disease): *Ars.*, *Cham.*, *Ipec*, *Merc.*, *Phos.*, *Phos. ac*, *Tart.*, *Verat.*

thin, watery, with nausea: *Ars.*, *Ipec.*, *Merc.*, *Phos.*

thin, watery, with vomiting of watery liquid and food: *Ars.*, *Cupr.*, *Ipec.*, *Phos.*, *Verat.*

NAUSEA AND VOMITING.

Nausea with thirst: *Bapt.*, *Bell.*, *Phos.*, *Verat.*

with gagging (retching): *Ant. tart.*, *Arn.*, *Asar. c.*, *Bell.*, *Bis.*, *Bry.*, *Chin.*, *Coloc.*, *Crot.*, *Hell.*, *Ign.*, *Ipec.*, *Jabor.*, *Nux v.*, *Pod.*, *Verat.*

Nausea with vertigo: *Camph., Merc., Verat.*

with continued pain at the pit of the stomach: *Acon., Ars., Bell., Camphor, Cham., Cupr., Merc, Nat. mur., Nux v., Phos., Pulsat., Rhus tox., Sulph., Tart., Verat.*

with diarrhœa: *Ars, Ipec., Merc., Phos.*

with hunger: *Ignat.*

with pale face and suppressed breathing: *Ipecac.*

Vomiting: *Acon., Ambra, Anac., Ant., Arg. n., Ars., Bell., Calc., Camph., Caust., Cham., China, Cic., Cin., Colc., Con., Cupr., Dig., Ferr., Gam., Grat., Ignat., Ipec., Iris v., Lach., Laur., Lyc., Nat. mur., Nux m., Nux v., Op, Petr., Phos., Plb., Pulsat, Sec. cor.. Sep., Sulph., Sulph. ac., Tart., Verat., Zinc.*

after a meal, with blueness of the lips: *Ars., Phos.*

after drinking: *Arn., Ars., Bry., Ipec., Nux vom., Puls., Verat.*

after drinking, with blueness of the face: *Ars., Verat.*

acrid: *Arg., Ferr., Hep., Iris, Ipecac.*

after drinking: *Arn., Ars., Bry., Nux, Puls., Verat*

foamy or frothy: *Æth., Crot., Cupr., Podo., Tart., Verat.*

Vomiting of food: *Ant , Ars , Æth., Bell., Bry.. Calc. carb., Carbo veg., Cham., Chin., Cina, Cocc., Colch , Coloc., Crot., Cupr., Ferr., Hyosc., Ignat., Ipec., Iris, Laur., Nat. mur., Nux, Phos., Phos. ac., Pod., Pulsat., Tart., Verat.*

of water, then food: *Nux vom.*

of food, then water: *Pulsat.*

of food, undigested: *Kali bich., Lyc., Phos.*

of watery liquid analagous to that of the stools, with pieces of mucus: *Ars., Bell., Camphor. Cuprum, Jatroph., Sec., Ipec , Stram., Verat.*

with pain in the stomach: *Ars., Bry., Camph., Cupr., Ipecac, Lach., Nux, Phos., Sulph., Stram., Tart., Verat.*

with colic: *Ars., Cupr., Nux, Phos., Puls., Stram., Tart., Verat.*

with colic and diarrhœa: *Ars., Cupr., Phos., Stram., Tart., Verat.*

with lassitude: *Ars., Camph., Ipecac, Phos., Verat.*

STOMACH.

Stomach, anxiety, distention and pressure at the pit of the: *Ars.*

burning in: *Ars., Bell., Bismuth, Bryo.,*

Camph., Cham., Cicuta, Carbo veg., Croton, Jatrop., Secale, Tabac., Verat.

Stomach, burning in, great: *Iris v.*

burning in the pit of: *Acon., Ars., Bell., Bry., Laur., Merc., Nux, Phos., Sec., Verat.*

burning heat in the pit of: *Ars., Camph., Hydro. ac , Phos.*

burning sensation in the, sometimes extending along the œsophagus to the mouth: *Ars.*

cramp in the: *Bell., Bry., Carbo, Cham., Cuprum, Nat. mur., Nux, Phos., Sec., Verat.*

coldness in: *Caps., Colch., Grat.*

continued pain at the pit, with nausea: *Acon., Ars., Bell., Camph., Cham., Cupr., Merc., Nat. mur., Nux, Phos., Rhus, Sulph., Tart., Verat.*

continued pain at the pit, with rumblings in the intestines: *Acon., Ars., Bell., Camph. Carbo veg., Cupr., Jat., Merc., Nat. mur , Nux, Phos., Phos. ac., Puls., Rhus, Secale, Sulph., Tart., Verat.*

pains in: *Arn., Ars., Coccul., Coloc., Cupr , Jat., Lyc., Zing.*

pains in, with vomiting and nausea: *Acon., Ars., Bell., Camph., Cham.,*

*Cupr., Ipec., Merc., Nat. mur., Nux.
Phos., Tart., Verat.*

Stomach, pressive or aching pains at the pit,
with liquid stools: *Ars., Camph., Cupr.,
Phos., Sec., Tart., Verat.*

pressive or aching pains at the pit, with
cramps or other spasms in the extremities or elsewhere: *Camph., Cupr.,
Phos., Phos. ac., Nat. mur., Sec., Tart.,
Verat.*

sensibility and swelling of the pit of:
Hep., Lyc., Sulph.

sensibility and swelling, with pains of
the extremities: *Ars., Camph., Cupr.,
Natr. m., Phos., Phos. ac., Verat.*

ABDOMEN.

Pains in the abdomen, with diarrhœa: *Ars.,
Cham., Ipecac, Laur., Merc., Merc. c.,
Natr. m., Nux, Phos., Rhus tox.,
Stram., Sulph., Tart., Verat.*

Rumblings in the intestines, with continued
pain in the pit of stomach: *Acon., Ars.,
Bell., Camph., Carb. v., Cupr., Jat.,
Merc., Natr. m., Nux, Phos., Phos. ac.,
Puls., Rhus, Sec., Sulph., Tart., Verat.*

in the intestines, with liquid stools: *Ars.,
Jatroph., Nux, Petr., Phos., Phos. ac.,
Puls., Rhus, Sec., Sulph., Tart., Verat.*

Throbbings in the abdomen: *Caps., Ignat., Op., Plumb., Sang., Tart.*

PULSE.

Pulse feeble and frequent: *Ars., Carb. v., Lach., Nux, Rhus, Rad.*

failing: *Ferr.*

full: *Acon., Bapt., Gels., Op.*

hard: *Acon., Æth., Bell., China.*

feeble and slow: *Camph., Dig., Laur., Merc., Puls., Rhus rad., Verat.*

feeble and small: *Ars., Camph., China, Dig., Laur., Nux, Phos. ac., Puls., Rhus., Verat.*

scarcely perceptible, with watery and white flocky stools: *Acon., Ars., Bry., Camph., Phos. ac., Rhus, Sec., Verat.*

imperceptible: *Ars., Carb. v., Crotal, Kali brom., Laur., Terb.*

intermitting: *Carbo veg., Hell., Nitric ac.*

irregular: *China, Laur., Tabac.*

rapid: *Acon., Æth., Ant. t., Ars., Bell., China, Jabor.*

slow: *Cupr., Dig., Laur., Mur. ac., Op.*

small: *Æth., Bell., Cupr.*

soft: *Bapt., Cupr.*

weak: *Ant. tart., Cupr., Dig., Kali brom., Kali c., Kreas., Merc. cor., Mur. ac., Tabac.*

HEART AND CHEST.

Heart, beating of, not rapid, but too violent: *Dig.*

    irregular action of: *Laur.*

    irregular action of, with great cardiac anguish: *Laur.*

    irregular action of, with suffocative attacks: *Laur.*

    oppression of: *Tabac.*

    palpitation of: *Ant. t.*, *Cact.*, *Cycl.*

    palpitation of, worse from least exertion: *Iod.*

Anguish in the chest: *Acon.*, *Ars.*, *Bell.*, *Bry.*, *Camph.*, *Carb. v.*, *Cicuta*, *Cupr.*, *Hydrocyan. ac.*, *Ipec.*, *Jatro.*, *Laur.*, *Nat. mur.*, *Phos.*, *Phos. ac.*, *Rhus tox.*, *Stram.*, *Verat.*

Constriction (spasmodic) of the chest: *Camph.*, *Caust.*, *Cupr.*, *Ferr.*, *Ipecac*, *Lach.*, *Nitric ac.*, *Nux*, *Op.*, *Phos.*, *Phos. ac.*, *Puls.*, *Spig.*, *Stram.*, *Sulph.*, *Verat.*

Cramps or tonic spasms in the chest: *Ars.*, *Bell.*, *Camph.*, *Caust.*, *Cic.*, *Cupr.*, *Fer.*, *Graph.*, *Hyos.*, *Ipecac*, *Kal.*, *Merc.*, *Nux*, *Op.*, *Phos.*, *Phos. ac.*, *Puls.*, *Sec.*, *Sep.*, *Stram.*, *Sulph.*, *Verat.*

Cramps in the muscles of the chest, with continual vomitings, and with the eyes turned upwards: *Camph.*, *Cic.*, *Verat.*

RESPIRATION.

Respiration difficult: *Arg. n.*, *Asaf.*, *Flat.*, *Puls.*

labored: *Apis*, *Arg. n.*, *Carb. v.*, *Cic.*, *Cupr.*

labored, with cold and blue skin: *Ars.*, *Camph* , *Carb. v.*, *Cupr.*, *Ipec.*, *Sec.*, *Verat.*

feeble: *China*, *Laur.*

oppressed: *Crotal.*, *Cupr.*, *Ipec.*, *Sulph.*, *Tabac.*, *Thuja*, *Verat.*

rattling: *Op.*

short: *Thuja.*

sighing: *Arg. n.*, *Ignat.*

slow: *Laur.*

snoring: *Op.*

VOICE.

Voice, choleraic: *Ferr.*

feeble: *Camph.*, *Sec.*, *Verat.*

hoarse: *Camph.*, *Carb. v.*, *Sec.*, *Verat.*

hoarse, face choleraic: *Ars.*, *Camph.*, *Carb. v.*, *Cupr.*, *Laur.*, *Phos.*, *Rhus*, *Sec.*, *Verat.*

hollow: *Sec.*

12

Voice, inaudible: *Sec.*
    last: *Carb. v.*
    weak: *Hell.*

Yawning: *Ant. tart., Elat., Plant., Podo.*

URINE.

Urine, retention of: *Canth., Lach., Op., Plum., Verat.*

    retention of, with ineffectual desire to urinate, at the commencement of reaction: *Canth., Verat.*

    scanty or suppressed: *Ars., Camph., Carb. v., Cupr., Ipec., Sec., Stram., Verat.*

    scanty, with fever: *Bell., Carb. v., Rhus t., Stram.*

    diminished, with cramps in the calves of the legs: *Ars., Carb. v., Cupr., Hyos., Lach., Lyc., Merc., Nux, Rhus, Secale, Sulph., Verat.*

    interrupted: *Con.*

    frequent: *Apis.*

    involuntary: *Bell., Caust , Cham., Hyos., Merc. v., Sep., Sil.*

PERSPIRATION.

Perspiration, cold: *Acon., Ars., Bell., Bry., Calc., Camph., Canth., Carb. v., Cham.,*

Chin., *Cin.*, *Cof.*, *Cupr.*, *Dulc.*, *Hell.*,
Hep., *Hyos.*, *Ignat.*, *Ipecac*, *Lach.*, *Lyc.*,
Merc., *Nat.*, *Nitricac.*, *Nux*, *Op.*, *Phos.*,
Phos. ac., *Plum.*, *Puls.*, *Rheum*, *Rhus*,
Sabad., *Sec.*, *Sep.*, *Sil.*, *Spig.*, *Stram.*,
Sulph., *Thuja*, *Tart.*, *Verat.*

Perspiration, viscid, clammy: *Ars.*, *Camph.*,
Ferr., *Hep.*, *Jat.*, *Lach.*, *Merc.*, *Nux*,
Phos., *Phos. ac.*, *Sec.*, *Verat.*

clammy, with slow pulse: *Verat.*

clammy, with spasmodic movements of
the jaw: *Camph.*, *Merc.*, *Nux*, *Phos.*,
Sec., *Verat.*

TONGUE.

Tongue, coldness of the: *Ars.*, *Bell.*, *Camph.*,
Laur., *Nat. mur.*, *Sec.*, *Verat.*

coldness of the, and breath: *Ars.*, *Carb.
veg.*, *Verat.*

coldness of the, with dryness: *Ars.*, *Bell.*,
Sec., *Verat.*

coldness of the, with cold sweat on the
body: *Ars.*, *Sec.*, *Verat.*

coated brown: *Bell.*, *Carb. v.*, *Hyos.*,
Rhus, *Sulph.*

coated white: *Ant. crud.*, *Ant. tart.*, *Arn.*,
Bell., *Bry.*, *Calc.*, *Carb.*, *Cupr.*, *Ignat.*,
Merc., *Nux*, *Petr.*, *Puls.*, *Sec.*, *Sulph.*

Tongue coated yellow: *Bell., Bry., Carb. v., Cham., Chin., Ipec., Nux, Puls., Verat.*

coated red: *Ars., Bell., Bry., Cham., Hyos., Lach., Nux v., Rhus, Sulph., Verat.*

coated red on the tip: *Rhus rad.*

coated red and yellow: *Bry., Cham., Nux, Verat.*

coated red and yellow, with pulse slow: *Bell., Rhus rad., Verat.*

coated black: *Ars., Lach., Merc. v.*

mapped: *Kali bich., Nat. mur.*

moist: *Bell., Phos.*

pale, reddish-blue: *Raph.*

rough: *Rhus tox.*

shining: *Apis, Lach.*

smooth: *Kali bich., Lach.*

swollen: *Merc. v.*

vesicles at the tip: *Lach.*

vesicles on the borders: *Apis.*

MIND AND MOOD.

Anguish: *Ars., Camph., Verat.*

Anxiety: *Acon., Canth., Carbo veg., Cic.*

Anger: *Ars.*

Apathy or indifference: *Ars., Bell., Cham., Camph., Cic., Hyos., Lach., Jatr., Op., Phos., Verat.*

Fear of Death, with internal burnings and tossing in bed: *Ars.*

Taciturnity, or repugnance to conversation: *Ars., Bell., Bry., Calc., Cham , Cic., Coloc., Cupr., Ignat., Lach., Merc., Nat. mur., Nux, Phos. ac., Puls., Sulph., Sulph ac., Verat.*

HEAD.

Confusion in the head: *Acon., Ars., Bell., Bry., Calc., Caust., Chin., Dig., Merc., Nux, Op., Phos. ac., Puls., Rhus., Sec., Verat.*

Heaviness or pressure in the head: *Acon., Arn., Ars., Bell., Bry., Calc., Camph., Carb. v., Cham., Chin., Cic , Ignat , Ipec., Lach., Laur., Lyc., Merc., Nat. mur., Nux, Op., Petr., Phos., Phos. ac., Puls., Rhus t., Sulph., Tart., Verat.*

Vertigo: *Acon., Ant., Arn., Ars., Bell., Bry., Calc., Camph., Carb. v., Caust., Cic., Cupr., Dig., Ferr., Graph., Hep., Hyos., Ignat., Ipec., Lach.. Laur., Lyc., Merc., Nat. mur , Nux, Op., Phos., Phos. ac., Puls., Rhus t., Thuja, Verat.*

with nausea and thirst: *Verat.*

Vertigo, with stupor: *Ars., Bell., Bry., Calc , Caust., Laur., Lyc., Merc., Nux, Op., Phos., Phos ac., Puls., Rhus t., Sec., Stram., Tart., Verat.*

EYES.

Eyes sunk in their orbits, with livid semi-circles under them: *Ars., Calc., Cic., Cupr., Kal., Laur., Phos., Phos. ac., Sec., Sulph., Verat.*

sunk in their orbits, with hoarse voice: *Ars., Calc., Camph., Cic., Cupr., Kal., Laur., Phos., Phos. ac., Sec., Sulph., Verat.*

upturned and fixed: *Camph., Cis., Verat.*

Pupils contracted : *Ars., Bell., Cham , Camph , Cic., Nux, Puls., Secale, Sep., Verat.*

bluish color about the: *Ars., Cupr., Phos. ac., Sec., Verat.*

FACE.

Face bluish: *Acon., Ars., Bell., Bry., Camph., Cham., Cic., Con., Cupr., Dig., Dros., Hep., Hyos., Ignat., Ipec., Lach., Lyc., Merc., Op., Phos., Puls., Samb., Spong., Staph., Stram , Tart., Verat.*

Face bluish and pale: *Ars., Bell., Bry., Camp., Cic., Con., Cupr., Dig., Dros., Hep., Hyos., Ignat., Ipec., Lach., Lyc., Merc., Op., Phos., Puls., Stram., Tart., Verat.*

Blueness of the lips: *Ang. spur., Ars., Camph., Caust., Berb., Chin. sulph., Cupr., Dig., Lyc., Phos.*

of the lips, with withered appearance of the skin: *Ars., Camph., Cupr., Lyc., Phos.*

under eyes; sleeps with eyes open: *Ipec., Phos. ac., Sulph., Verat.*

of the skin and face: *Ars., Cupr., Verat.*

of the skin and face, with slow pulse and clammy sweat: *Verat.*

Face choleraic: *Ars., Cupr., Carb. v., Ipec, Laur., Phos. ac., Rhus, Sec., Verat.*

Spasm of the jaw: *Bell., Cham., Cic., Cupr. ac., Hydro., Lach., Laur., Op., Rhus, Sec., Verat.*

UPPER EXTREMITIES.

Cramps in the upper arms: *Ac., Phos., Sec.*

in the forearms: *Laur., Phos. ac., Sec.*

in the wrist: *Phos. ac.*

Coldness of the hands: *Acon , Bell., Cham., Ipec., Natr. mur., Nux, Petr., Phos., Sulph., Tart., Verat.*

Cramps in the hands: *Bell.*, *Calc.*, *Coloc.*, *Graph.*, *Laur.*, *Phos. ac.*, *Sec.*, *Stram.*

of the fingers: *Ars. Arn.*, *Lyc.*, *Nux*, *Phos.*, *Phos. ac.*, *Sec.*, *Sulph.*, *Verat.*

of the fingers, with clammy perspiration: *Ars*, *Ferr.*, *Nux*, *Verat.*

LOWER EXTREMITIES.

Coldness of the feet: *Acon.*, *Bell.*, *Calc.*, *Caust.*, *Dig.*, *Ipec.*, *Lach.*, *Lyc.*, *Merc.*, *Phos.*, *Plat.*, *Plumb.*, *Rhod.*, *Rhus t.*, *Sulph.*, *Tart.*, *Verat.*

in the hips: *Coloc.*, *Phos. ac.*

in the thighs: *Camph.*, *Cann.*, *Hyos.*, *Ipec.*, *Merc.*, *Phos. ac.*, *Rhus*, *Sep.*, *Verat.*

in the hands: *Calc.*, *Phos.*

in the legs: *Carbo v.*, *Coloc.*, *Cupr.*, *Jat.*, *Phos. ac.*

in the calves of the legs: *Ars.*, *Bry.*, *Calc.*, *Camph.*, *Carb. v.*, *Cham.*, *Coff.*, *Coloc.*, *Cupr.*, *Hyos.*, *Jat.*, *Lach.*, *Lyc.*, *Merc. ac.*, *Nit.*, *Nux*, *Phos.*, *Rhus*, *Sec.*, *Sal. n.*, *Sulph.*, *Tart.*, *Verat.*

in the calves of the legs, with burning heat in the stomach: *Ars.*, *Camph*, *Phos.*

Coldness in the calves of the legs, with diminished secretion of urine: *Ars.*, *Carb. veg.*, *Cupr.*, *Hyos.*, *Lach.*, *Lyc.*, *Kali bich.*, *Merc.*, *Nux*, *Rhus*, *Sec.*, *Sulph.*, *Verat.*

in the calves, coldness of the feet: *Lach.*, *Lyc.*, *Phos.*, *Rhus t.*, *Carbo veg.*, *Verat.*

in the feet: *Camph.*, *Caust.*, *Graph.*, *Lyc.*, *Nux*, *Sec.*, *Tart.*, *Verat.*

in the feet, with burning in the stomach: *Camph.*

in the soles of feet: *Calc.*, *Carb. v.*, *Coff.*, *Fur.*, *Hep.*, *Phos. ac.*, *Plumb.*, *Sec.*, *Sulph.*

in the toes: *Calc.*, *Fur.*, *Hep.*, *Merc.*, *Nux*, *Sec.*, *Sulph.*

SKIN.

Coldness of skin: *Acon.*, *Ant.*, *Arn.*, *Ars*, *Bell.*, *Bry.*, *Camph.*, *Canth.*, *Carb. v.*, *Caust.*, *Cham.*, *Chin.*, *Cic.*, *Cupr.*, *Dros.*, *Dulc.*, *Fur.*, *Graph.*, *Hyos.*, *Ipec.*, *Lach.*, *Laur.*, *Lyc.*, *Merc.*, *Nux*, *Op.*, *Phos.*, *Phos. ac.*, *Rhus*, *Sabad.*, *Sec*, *Stram.*, *Sulph.*, *Verat.*

Blueness of the skin: *Acon.*, *Ars.*, *Bell.*, *Bry.*, *Carb. v.*, *Cupr.*, *Lach.*, *Nux*, *Op.*, *Rhus*, *Sec.*, *Verat.*

13

Cold and bluish, with cold perspiration: *Ars* , *Caust.*, *Cupr.*, *Secale*, *Verat.*

Blueness of the skin in different parts, and withered appearance: *Cupr.*, *Secale*, *Verat.*

FEVER.

Chill: *Camph.*, *Dig.*
mingled with heat: *Dig.*

Chilliness: *Arg. n.*, *Asar. e.*, *Camph.*, *Cicuta*, *Dig.*, *Elat.*, *Merc.*, *Puls* , *Sulph.*
when leaving the fire: *Aloe.*

Coldness: *Æth.*, *Camph.*, *Jat.*, *Laur.*, *Tabac.*

Shuddering: *Acon.*, *Camph.*
internal: *Acon.*
without coldness: *Lach.*

INDEX.

www.ingramcontent.com/pod-product-compliance
Lightning Source LLC
Chambersburg PA
CBHW021945190326
41519CB00009B/1144